ANIMALS

Ugly Bugs

Kerri O'Donnell

PowerKiDS
press

New York

Published in 2007 by The Rosen Publishing Group, Inc.
29 East 21st Street, New York, NY 10010

Book Design: Michael Ruberto

Photo Credits: Cover (left), p. 3 © Garth Helms/Shutterstock; cover (top right), p. 15 © Ra'id Khalil/Shutterstock; cover (bottom right) © Alexander M. Omelko/Shutterstock; p. 5 © H. Tuller/Shutterstock; p. 7 © Anthony Bannister; Gallo Images/Corbis; p. 9 © Gordana Sermek/Shutterstock; p. 11 © M. Fogden/OSF/Animals Animals; p. 13 © Frank B. Yuwono/Shutterstock; p. 17 © Joy Stein/Shutterstock; p. 19 © Susan Ellis/www.insectimages.org; p. 21 © Richard T. Nowitz/Corbis; p. 22 © Nicolas Raymond/Shutterstock.

Library of Congress Cataloging-in-Publication Data

O'Donnell, Kerri, 1972-
  Ugly bugs / Kerri O'Donnell.
      p. cm. - (Ugly animals)
  Includes bibliographical references and index.
  ISBN-13: 978-1-4042-3527-2
  ISBN-10: 1-4042-3527-2
  1. Insects-Juvenile literature. I. Title.
  QL467.2O36 2007
  595.7-dc22
                              2006014622

Manufactured in the United States of America

Contents

Ugly Bugs

You've seen plenty of bugs—like bees, ants, and butterflies. Bugs live almost everywhere. We know about 1 million kinds of bugs. Every year, scientists discover thousands more kinds. That's a lot of bugs!

Bugs, or **insects**, might look different from each other. However, all bugs have six legs and three main body parts. Some bugs are tiny. Others are much larger. Some bugs have beautiful colors. Others are just plain ugly. Let's look at some ugly bugs!

You can find bugs outside or even in your house!

Creepy Cockroaches

Do cockroaches give you the creeps? There are about 3,500 kinds of cockroaches. The cockroaches we usually see are household pests. These cockroaches live in warm, dark places like the cracks in walls or behind stoves. Once they get into a building, they are hard to get rid of.

Cockroaches eat anything they find. If they can't find food, they might eat soap, paper, or glue. They even eat their own skins or other cockroaches. Gross!

A cockroach can live for a while without its head!

Stag Beetles

Stag beetles are very strange, scary-looking bugs. Some male stag beetles have huge **jaws** that look like the horns of a male deer, or stag. This is how the beetle got its name.

Sometimes a stag beetle's jaws are almost as long as the beetle's whole body. In the southern United States, the giant stag beetle has jaws that are 1 inch (2.5 cm) long!

Male stag beetles sometimes use their large jaws for fighting.

Peanut-Head Bugs

The peanut-head bug lives in the forests of Central America and northern South America. It gets its name because its strange head is long and shaped like a peanut shell. This "peanut" is actually a really big forehead. The bug's eyes are found just above its front legs.

Part of the peanut-head bug's stomach is found inside the "peanut" on its head! Peanut-head bugs eat tree sap. The sap is stored in this part of the stomach.

A peanut-head bug can give off a bad smell if an enemy comes too close.

The Praying Mantis

The creepy-looking praying mantis got its name because it holds up its front legs like it is praying. The praying mantis has a triangle-shaped head with a big eye on each side. Its body often matches the color and shape of plants and leaves. This helps it hide from danger.

A praying mantis uses its front legs to grab its **prey**. Sharp hooks on the legs help the praying mantis hold on to its victim. Ouch!

A female praying mantis will sometimes eat a male praying mantis!

A Bug or a Stick?

Have you ever seen a twig on the ground that got up and walked away? That was a walkingstick!

Walkingsticks look exactly like their name sounds—they look like sticks that walk. This helps them hide from enemies. When they are resting, they look just like brown or green twigs. Walkingsticks are different from most insects because they have no wings and cannot fly. Their long, skinny legs help them move along leaves and branches.

When a young walkingstick loses one of its legs, it can grow a new one!

Dung Beetles

Dung beetles get their food from the waste—or dung—of large, plant-eating animals like cows. The dung is made of water and broken-down plants the animals have eaten. Dung beetles squeeze the dung in their mouths and drink its juice. It may sound gross, but the juice has lots of **nutrients** in it. The dung beetles then lay their eggs in the dung. The eggs get nutrients from the dung, too.

A dung beetle makes a dung ball. The dung ball will give the beetle's eggs food and shelter.

Stinky Stink Bugs

Stink bugs are not only creepy to look at—they stink, too! Stink bugs got their name because they smell bad. When they are in danger, they let off a stinky liquid. The stink bug's enemies will then usually leave it alone. They don't like the gross smell!

Some stink bugs are pests to farmers. They eat plants, fruits, and vegetables, and can ruin farmers' crops. If stink bugs get into your house, they can make everything they touch stink!

The red circles show where the stinky liquid comes out of the stink bug's body.

Bloodsucking Mosquitoes

You've probably been bitten by a mosquito. Some female mosquitoes feed on human blood. The mosquito uses a long, tube-shaped body part to cut the skin and suck out blood. The mosquito's body then swells up with blood.

Most mosquitoes we see just give us an itchy bump with their bites. However, these nasty bugs can also spread deadly **diseases** with their bites. In some parts of the world, these diseases kill many people each year.

This female mosquito is sucking a person's blood.

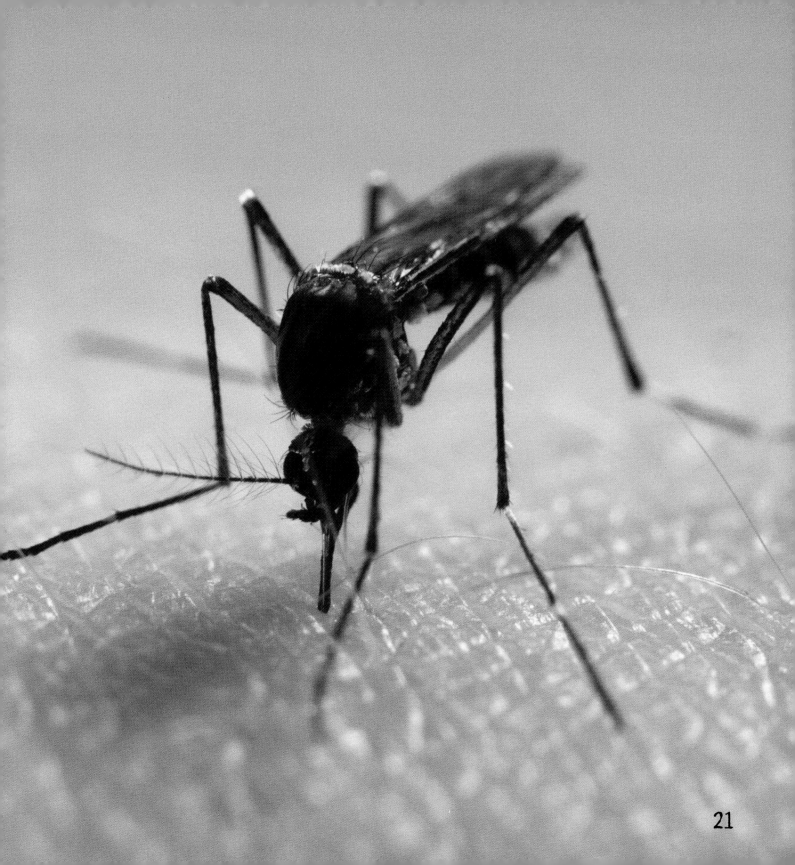

21

Bugging Out!

We've learned about some truly ugly bugs. Some bugs—like stink bugs and mosquitoes—can cause harm. Others can help us, even if they look weird or have weird habits. The praying mantis eats insects that can harm farmers' crops. Dung beetles dig tunnels in the ground and help keep the soil full of nutrients so plants can grow.

The next time you see an ugly bug, remember that the creepy critter might be helping you!

Glossary

disease (dih-ZEEZ) A sickness.

insect (IHN-sekt) A small animal without bones that has three main body parts, six legs, and usually has wings.

jaw (JAW) The upper or lower parts of the mouth.

nutrient (NOO-tree-uhnt) Anything that is needed by living things to live and grow.

prey (PRAY) An animal that is hunted as food by another animal.

Index

Web Sites

Due to the changing nature of Internet links, PowerKids Press has developed an online list of Web sites related to the subject of this book. This site is updated regularly. Please use this link to access the list:
http://www.powerkidslinks.com/uglyani/bugs/